JOBS KIDS WANT

WHAT'S IT REALLY LIKE TO BE A CODER?

CHRISTINE HONDERS

PowerKiDS press.
New York

Published in 2020 by The Rosen Publishing Group, Inc.
29 East 21st Street, New York, NY 10010

First Edition

Editor: Greg Roza
Book Design: Michael Flynn

Photo Credits: Cover, pp. 1, 13 Hero Images/Getty Images; pp. 4, 6, 8, 10, 12, 14, 16, 18, 20, 22 (background) Apostrophe/Shutterstock.com; p. 5 Ekaterina Vidyasova/Shutterstock.com; p. 7 Rawpixel.com/Shutterstock.com; p. 9 Shahrokost/Shutterstock.com; p. 11 https://en.wikipedia.org/wiki/Ada_Lovelace#/media/File:Ada_Lovelace_portrait.jpg; pp. 15, 19 REDPIXEL.PL/Shutterstock.com; p. 17 Altitude Visual/Shutterstock.com; p. 21 Gorodenkoff/Shutterstock.com; p. 22 Africa Studio/Shutterstock.com.

Cataloging-in-Publication Data

Names: Honders, Christine.
Title: What's it really like to be a coder? / Christine Honders.
Description: New York : PowerKids Press, 2020. | Series: Jobs kids want | Includes glossary and index.
Identifiers: ISBN 9781725301689 (pbk.) | ISBN 9781725301702 (library bound) | ISBN 9781725301696 (6 pack)
Subjects: LCSH: Computer programming–Vocational guidance–Juvenile literature.
Classification: LCC QA76.25 H66 2020 | DDC 005.1023–dc23

Manufactured in the United States of America

CPSIA Compliance Information: Batch #CSPK19. For Further Information contact Rosen Publishing, New York, New York at 1-800-237-9932.

CONTENTS

Coding in Our Lives

Computers are everywhere. We use them at school and at home. When you play your favorite video game, have you ever wondered how it works? Computers can't do these things by themselves. Coders tell computers what to do.

What Is Code?

Code is a set of instructions that computers can understand. Computers can only read a language made up of 0s and 1s. It would take forever for coders to write code using 0s and 1s. So, they learn to code in a language that's easier to read and write.

Programming Languages

Coders give computers directions with coding languages, or **programming languages**. Then, the commands are explained to the computer in a language it understands. Java is the most popular programming language in the world. Python is used by the Internet company Google.

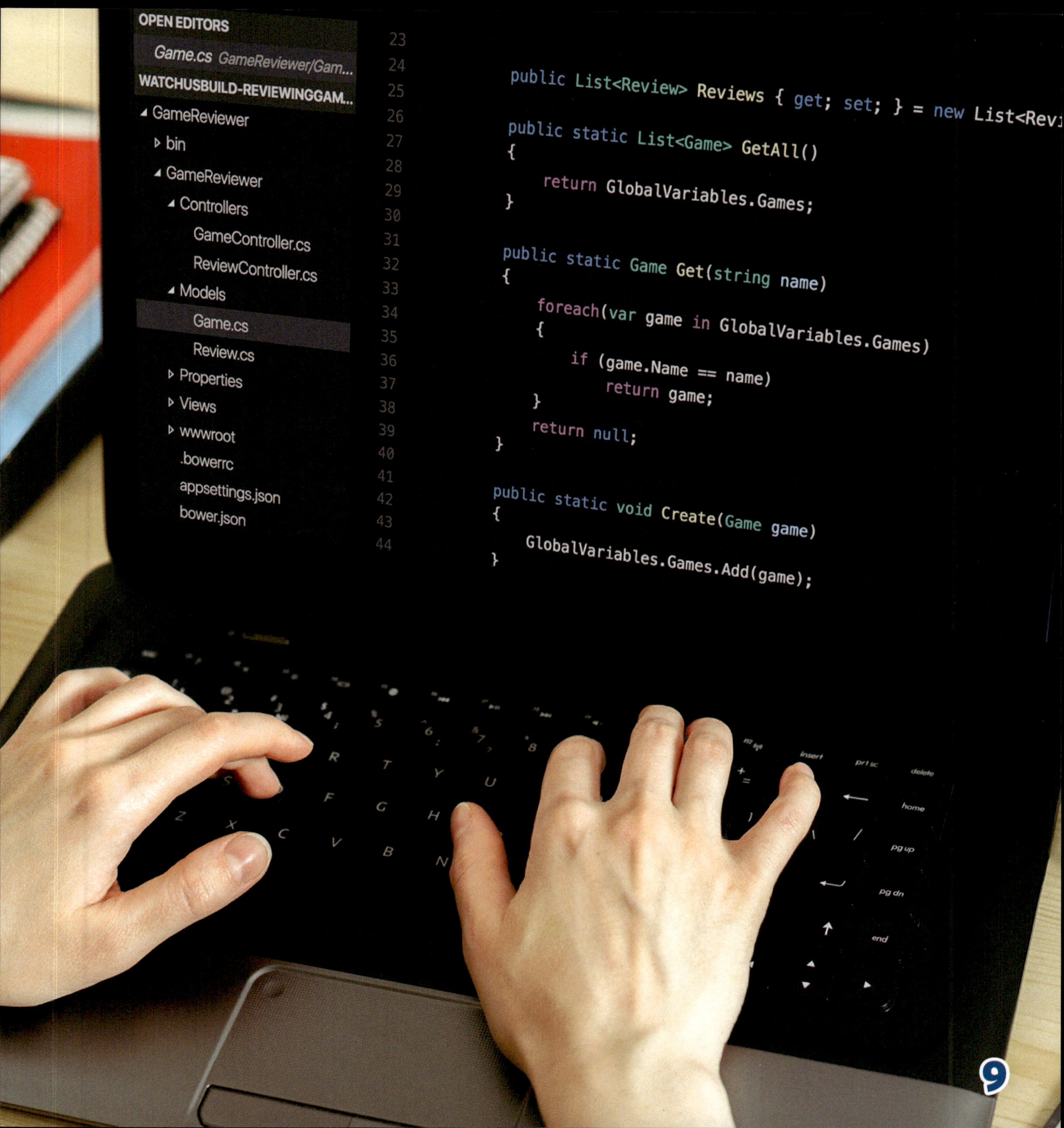
OPEN EDITORS
Game.cs GameReviewer/Gam...
WATCHUSBUILD-REVIEWINGGAM...
GameReviewer
bin
GameReviewer
Controllers
GameController.cs
ReviewController.cs
Models
Game.cs
Review.cs
Properties
Views
wwwroot
.bowerrc
appsettings.json
bower.json
public List<Review> Reviews { get; set; } = new List<Rev
public static List<Game> GetAll()
{
return GlobalVariables.Games;
}
public static Game Get(string name)
{
foreach(var game in GlobalVariables.Games)
{
if (game.Name == name)
return game;
}
return null;
}
public static void Create(Game game)
{
GlobalVariables.Games.Add(game);
}

First Coder

The first computer program was written in 1843. The coder who wrote the program was an English woman named Ada Lovelace. She wrote the program for a friend who created a machine with moving parts. That machine was an early type of computer.

How to Become a Coder

Coders study math and computer science in school. Some coders go to college. Coders that go to college can usually get better jobs. There are many places on the web that teach coding. Some coders use them to learn how to code.

CODE
CAMP

What Do Coders Do?

Coders make **software** that's used on computers and tablets. They create websites and apps for smartphones. Once coders build something, they make sure it works. They test it for mistakes, called bugs. They create **updates**, so the program keeps running smoothly.

18:00
CALL

The Good and the Bad

Coding is a creative way to make your ideas come to life. Coders have many job choices and can make good money. But coding can be tough. Coders sit for a long time staring at computers. This hurts their eyes and backs. Hours of typing can cause wrist **injuries**.

Coding Skills

Coders write programs for others, so they must be good listeners. They often work in teams. Code takes a long time to write, so **patience** is important. Coders must pay close attention to their work. One tiny mistake can ruin the whole program!

Learn Coding Now!

It's never too early to learn coding. It helps your brain solve problems faster. You'll learn to use computers to **express** yourself in a new way. Coding lets you share your ideas with the world. It's also a lot of fun!

The Future of Coding

In ten years, there will be one million jobs for coders. We need coders to fill those jobs and keep our computers going. If you love computers, you should learn coding. One day, you may know what it's really like to be a coder!

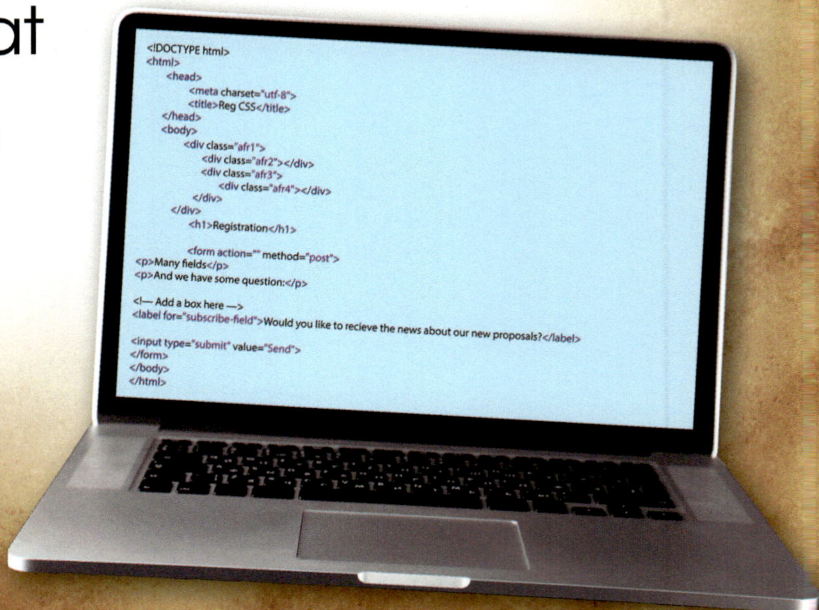

GLOSSARY

express: To share your thoughts, feelings, and interests with others.

injury: Hurt, damage, or loss.

patience: The ability to remain calm when dealing with something difficult.

programming language: A set of rules that explain how code should be written.

software: Programs that run on computers and perform certain jobs.

update: An up-to-date form of something, such as a computer program.

INDEX

WEBSITES

Due to the changing nature of Internet links, PowerKids Press has developed an online list of websites related to the subject of this book. This site is updated regularly. Please use this link to access the list: www.powerkidslinks.com/JKW/coder